Beyond Romanticism

Questioning the Green Gospel

Samuel Gregg

2000

Beyond Romanticism

Questioning the Green Gospel

Samuel Gregg

CIS Occasional Papers 73

Published July 2000 by

The Centre for Independent Studies Limited
PO Box 92, St Leonards, NSW 1590
Email: cis@cis.org.au
Website: **www.cis.org.au**

National Library of Australia
Cataloguing-in-Publication Data:

Gregg, Samuel, 1969- .
Beyond romanticism : questioning the green gospel.

Includes index.
ISBN 1 86432 051 6.

1. Ecology - Religious aspects - Christianity. I. Title.
(Series : CIS occasional papers ; 73).

261.8362

Cover and book design by Heng-Chai Lim
Edited by Susan Windybank
Printed by Australian Print Group, Victoria
Typeset in Garamond 11pt

Contents

Foreword

Over the past forty years, environmentalism has emerged as one of the major political movements that transcend national boundaries. Its concerns have motivated people from a variety of backgrounds to become politically active and have significantly altered the manner in which the state as well as key institutions of civil society such as business thinks about environmental issues. The churches have not been unaffected by this phenomenon. Many Christians, both lay and clerical, are to be found at the forefront of environmentalist lobbying.

In this Occasional Paper, Samuel Gregg raises questions about the adequacy of many Christian pronouncements and commentaries on environmental issues. Too often, he contends, they are characterised by questionable theological premises as well as a tendency to accept uncritically arguments articulated by the green lobby.

In making these points, however, Gregg does not focus solely upon critiquing those Christian contributions which tend to overrate (sometimes grossly) the environment's significance in the Christian vision of the world (at times, he claims, these verge on the pantheistic). Instead, his primary concern is to clarify how Christians should reflect upon and approach environmental questions in a manner consistent with basic Christian doctrines. In doing so, he demonstrates that while Christians cannot view the natural world as simply something to be exploited, they can easily harmonise respect for nature with the belief that the natural world and all it contains may be legitimately used by individuals and business to serve human needs.

Greg Lindsay
Executive Director
The Centre for Independent Studies

About the Author

Samuel Gregg has an MA in political philosophy from the University of Melbourne and a Doctorate of Philosophy in moral philosophy from the University of Oxford. The author of *Challenging the Modern World: Karol Wojtyla/John Paul II and the Development of Catholic Social Teaching* (1999) as well as many monographs and occasional papers, he is Resident Scholar at The Centre for Independent Studies.

Beyond Romanticism
Questioning the Green Gospel

> Man is created to praise, reverence and serve God our Lord, and by this means to save his soul. The other things on the face of the earth are created for man to help him in attaining the end for which he is created. Hence, man is to make use of them in as far as they help him in the attainment of his end.
>
> St Ignatius of Loyola

Introduction[1]

In his famous biography of St Francis of Assisi—the patron saint of those who promote ecology (John Paul II 1979)—G.K. Chesterton went to some length to point out that though he loved nature, Francis, unlike pre-Christian pagans, never worshipped nature itself ([1923] 1987).

In more recent times, however, some Christian thinkers appear to have blurred the distinction between respecting nature and worshipping it. Matthew Fox, an ex-Catholic priest, depicts the earth in his book, *The Coming of the Cosmic Christ*, as a Christ-like figure. He even describes the orthodox Christian view that the person of Jesus Christ is Revelation as 'Christofascism' (1988: 164). The underlying theme of Fox's book is to urge the churches to move 'beyond' a theology of sin and redemption and develop a theology in which nature itself is Revelation.

The Christian churches have not escaped the impact of widespread concern about the environment that has emerged

---

[1] The author would like to thank Professor Ron Duncan, Professor Ian Harper, Professor Helen Hughes, Professor Eric Jones, Rev. Dr Rodger Charles, S.J., Dr Jo Kwong, Barry Maley, and Susan Windybank for their comments on this text. Remaining errors are the author's responsibility.

over the past forty years. Environmentalism's impact upon Western public policy and social thinking in this time has been considerable. Not only do governments devote much time to explaining how they will address environmental issues, but businesses and other institutions of civil society also find themselves having to clarify their position on 'green' questions. Likewise, many church leaders and thinkers now spend much time discussing and writing about environmental issues. In 1999, for example, the church-based National Partnership for the Environment was founded in the United States. Located in New York, this organisation describes itself as seeking to make environmentalism a central element of church and synagogue life. The Partnership's director, Paul Gorman, states that it wants to ensure that

> the next generation of religious leaders hold care for creation as a defining vocation and ministry. This is, for me personally and for many others, a profoundly prophetic vision. It goes to the heart of what religious life must mean. It brings into question the most fundamental tenets and teachings of our traditions. (Gorman quoted in Tooley 1999: 8)

One need not be a theologian to discern the problematic notions underlying these statements. Should care for the environment really be regarded as a *defining* vocation for future religious leaders? Surely for Christians, the preaching of the message of Jesus Christ is definitive of every Christian's role. One would also hope that Gorman is not seriously suggesting that basic Christian doctrines should be substantially 'revised' or even discarded, depending upon the degree to which they allegedly contribute to environmental problems. This would lead to the remarkable conclusion that Christian doctrine somehow has to prove its continuing validity before the tribunal of environmental orthodoxy.

It would be easy to dismiss the citation above as a somewhat emotivist statement made by the director of one church-associated organisation. Subtle but also questionable incursions of environmentalist thinking and priorities have, however, manifested themselves in the writings of a variety of other Christian thinkers.

Our purpose, then, is not to engage in discussion of environmental issues *per se*. Nor does this paper attempt to contribute to ongoing debates about how to protect the environment without unduly impeding economic development. Instead, the intention is to clarify how Christians may approach environmental questions in a manner consistent with basic Christian doctrines. In doing so, it hopes to help forestall attempts on the part of those Christians (or non-Christians for that matter) who might be tempted, unwittingly or otherwise, to turn churches into highly-politicised environmental lobby groups and distract them from their central task of leading people to salvation. This is not insignificant if the American theologian Michael Novak is correct when he states that '[o]ne can predict with some certainty that environmentalism is likely to replace Marxism as the main carrier of gnosticism (and anti-capitalism) in the near future' ([1982] 1991: 435).

The possibility that some Christians could venture down this path is not as far-fetched as some might suppose. A perennial problem that has manifested itself in the Christian churches, especially in the twentieth century, has been the tendency of some Christians to fall slowly into the trap of subordinating Christian doctrine to the principles and agendas of particular ideologies or transitory intellectual fashions.

Naturally, Christians must be attentive to what is happening in the world. Yet this does not mean that their reflection upon secular developments should be unthoughtful. One of the problems, for example, characterising the writings of some scholars who embraced the theologies of liberation in the late 1960s was the extent to which many uncritically accepted ideas derived from thinkers such as Karl Marx, Theodore Adorno, Max Horkheimer, Jurgen Habermas and Herbert Marcuse as well as the 'New Left' movements in Western Europe and Latin America (Ratzinger 1985: 177-8; Gregg 1999a: 193-4, 206).

It is consequently important that Christians concerned about the environment—but also determined to remain faithful to the

Gospel message—should be clear about which biblical, theological and philosophical frameworks best allow them to contribute to such debates without diluting their Christian beliefs or finding themselves lapsing into mild versions of pantheism.[2] This paper seeks to contribute to the formation of such frameworks in four ways:

- it outlines some of modern environmentalism's primary philosophical underpinnings and demonstrates how they have manifested themselves in the writings of some contemporary Christian thinkers;
- it asks what one of the most important texts of Scripture, the Book of Genesis, suggests to Christians about how they should view the world of nature;
- it applies this framework to a contemporary issue promoted by many environmentalists, this being animal 'rights', so as to demonstrate how such a vision should affect Christians' understanding of particular environmental matters; and
- it highlights two areas where the Christian churches may be in a position to raise serious questions about environmentalism's adequacy as a worldview.

The roots of environmentalism

In 1999, a paper produced by Stephen Moore, based on the findings of the recently deceased economist Julian Simon, pointed out that

---

[2] Broadly speaking, pantheism holds that God and the world are one. Taken in its strictest sense—i.e. identifying God with the world—pantheism is simply pure materialism. One should not be surprised, then, that Christians view it as a heresy. The Catholic Church, for example, has repeatedly condemned pantheism as an error. Pius IX's *Syllabus Errorum* (1864), for example, condemned the proposition that: 'There is no supreme, all-wise and all-provident Divine Being distinct from the universe; God is one with nature and therefore subject to change; He becomes God in man and the world; all things are God and have His substance; God is identical with the world, spirit with matter, necessity with freedom, truth with falsity, good with evil, justice with injustice' (Denzinger [1957] 1998: para. 1701). Likewise, the First Vatican Council anathematises those who assert that the substance or essence of God and of all things is one and the same (Denzinger [1957] 1998: para 1803).

> There is almost certainly no other issue about which . . . general preconceptions are so contrary to objective reality as they are about the environment. Most . . . believe that because of industrialization, population growth, and our mass-consumption society, the quality of our air and water is deteriorating and that our natural resources will soon run dry. The scientific evidence tells us exactly the opposite . . . (Moore & Simon 1999: 28)

Unfortunately, perceptions rather than evidence tend to shape the character of much political discussion. Few would question that widespread perceptions of increasing environmental degradation have led to greater public pressures for greater protection of the environment.

But when it comes to understanding the *philosophical* origins of modern environmentalism, our understanding of this phenomenon is likely to become more complex. It is reasonable to suggest, for example, that some roots of modern expressions of environmentalism may be traced back to the eighteenth century French Enlightenment thinker, Jean-Jacques Rousseau, not least the essay that first brought him fame, his *Discourse on Inequality* (1749).

In this relatively short piece, Rousseau suggested that, contrary to his contemporaries' view, human manners and morals had been corrupted by the advancement of the arts and sciences. Rousseau's fundamental thesis was that those who claimed that history illustrated humanity's progressive development from a barbarous 'state of nature' towards a more civilised society were wrong. The state of nature was, as imagined by Rousseau, not at all barbarous. Instead, it was a world in which humans were simple animal-like creatures. Far from being fallen beings (as Christianity holds), Rousseau claimed that primitive man lived happily in an unreflective state of pure being. He spent his time meandering 'up and down forests, without industry, without speech, and without home, an equal stranger to war and to all ties, neither standing in need of his fellow creatures nor having any desire to hurt them, and perhaps even not distinguishing them from one another' (Rousseau [1755] 1997: 79).

The transition from this existence to economic and civil society was, according to Rousseau, a terrible loss. As humans invented agriculture, and then engaged in trade and commerce, they apparently lost touch with the natural world and came to depend upon one another. One of the worst developments, Rousseau maintained, was when people began to enclose pieces of ground, claimed them as their own, and persuaded others to believe in the validity of their claim:

> The first man who, having enclosed a piece of ground, bethought of himself saying *This is mine*, and found people simple enough to believe him, was the real founder of civil society. From how many crimes, wars and murders, from how many horrors and misfortunes might not anyone have saved mankind, by pulling up the stakes, or filling up the ditch, and crying to his fellows: 'Beware of listening to this imposter; you are undone if you once forget that the fruits of the earth belong to us all, and the earth itself to nobody'. (Rousseau [1755] 1997: 76)

Rousseau's thoughts on these and related matters have encountered no shortage of critics, both religious and secular (Maritain 1970; Hayek 1988: 50). It remains, however, that Rousseau's evocative fantasies about primitive society contributed to nineteenth century Romanticism's extolling of the simple life as well as the rise of the romantic literary genre that celebrates the primacy of feeling and the beauties of nature, not to mention the neo-pagan cult of nature that underlay much National Socialist ideology (Glendon 1999: 44, 47)—something rarely mentioned or perhaps even known by most environmental activists today.

To a certain extent, however, Rousseau's ideas do prefigure much unease on the part of modern man about humanity's relationship with the natural world. After the calamities of the twentieth century, one is entitled to think that unless progress, technological or otherwise, occurs within a particular moral-cultural framework, then the consequences can be disastrous

(CDF 1986: paras. 5-19, 21). One of the characteristics of modernity has been the progressive mastery that humans have achieved over nature. Many people no longer have as much contact as their ancestors did with nature in its primitive form. Rather, they are more regularly in touch with natural forces that have been recomposed by human ingenuity. According to the patristic scholar Jean Daniélou, S.J., this has induced a certain degree of anguish within man about his own powers (Daniélou 1961: 122), not least because of the extent to which such progress has enhanced humanity's capacity to destroy itself.

Environmentalism's emergence can be seen on one level as an appropriate corrective to tendencies within modern society to view its technological and material progress as somehow providing its own justification—just as humanism, insofar as it asserts the inherent worth of human life, may be considered a corrective to some Christians' tendency to see life in this world as nothing but a means to reach the next. But there is surely much truth in David Elder's statement that we seem to have moved from a situation of proclaiming the might of technology to the opposite extreme of underlining the fragility, even the 'sacredness' of the planet (1996: 127-128).

In its most extravagant form, such trends are typified by writings of the 'Deep Ecologists'. Many Deep Ecologists preach a type of absolutist biological egalitarianism, in which all species, animal or plant, are considered equals. Such thinking also manifests itself in the writings of those who adhere to the idea of 'Gaia', a concept first advanced by scientist James Lovelock (1974). Its basic premise is that the Earth itself is a living superorganism, and some Gaia enthusiasts come close to attributing it with divine status.

One characteristic shared by this more radical stream of environmental thought is a tendency to view the environment as a placid, harmless, semi-paradisal Panglossian world. In this Rousseauian-like vision, nature is considered to embody a self-regulating harmony. It is also considered naturally hospitable to man—provided that humans leave it alone.

Upon sober reflection, one soon realises the naïvety of such views. Much destruction has been wrought upon humanity by nature, be it in the form of natural disasters or virulent diseases. But with or without man's presence, the natural world remains a far from harmonious paradise. Unpredictable things happen. Winds erode land. Storms wash away entire territories. Earthquakes cause havoc. Animals are hardly kind to each other. Millions of species have become extinct without humans playing any part in their demise. It is therefore important that when confronted with any form of environmental utopianism, Christians—indeed anyone of good will—should recall the words of the American theologian Robert Royal:

> Nature itself and the God who created it show little in common with today's usual environmental view of the world as a constant, benign and nurturing place except when foolish or outright evil humans disrupt it . . . The notion of nature as our Mother, as a being greater and better than ourselves, is—and always was—mistaken . . . Our concern for nature has to acknowledge the imperfection of the world even as it recalls the biblical assurance that creation is good. (Royal 1999: 132)

Another underlying theme of much environmentalist theory is a generally negative view of free enterprise and economic development. An example of such thought may be found in the writings of Rudolf Bahro, one of the leading theoreticians of the German Greens. Bahro refers to the 'clearly and markedly self-destructive, outwardly murderous and inwardly suicidal character of our industrial civilization . . . the simultaneously most expansive (aggressive) and most effective (productive) economic system in world history, the capitalist mode of production'. He then claims that the industrialisation driven by business entrepreneurship not only destroys its own preconditions for existence, i.e. natural resources, but 'also the natural foundations of human life, of the very biosphere that sustains us. The completion of this process on a world scale would be the ultimate natural catastrophe' (Bahro 1986: 11-12).

Such a view of industrialisation contrasts sharply with Paul VI's interpretation in his social encyclical *Populorum Progressio* (1967). Though Pope Paul acknowledged that industrialisation had its negative aspects and expressed concerns about the environment (1971), he nonetheless insisted:

> The introduction of industrialization, which is necessary for economic growth and human progress, is both a sign of development and a spur to it. By dint of intelligent thought and hard work, man gradually uncovers the hidden laws of nature and learns to make better use of natural resources. As he takes control over his way of life, he is stimulated to undertake new investigations and fresh discoveries, to take prudent risks and launch new ventures, to act responsibly and give of himself unselfishly. (Paul VI [1967] 1971: para. 25)

To this, one could add the observation that humanity's increasing mastery over the forces of nature through its work and use of technology has assisted in purifying religious belief. According to Daniélou, technology

> *frees religion and the supernatural from a whole cumbersome burden of the pseudo-supernatural and the pseudo-religious.* Primitive man identified the supernatural everywhere, but largely on account of his ignorance. Purification of genuine religion from such degradations results from man's investigation of the whole range of his powers. There is therefore, in this sense, a wholly positive contribution from the technological world to the religious world. (1961: 123-4)

Greening the gospel

In recent years, there has been a certain tempering of excessive claims made by some environmental thinkers. Left-liberal environmental authors such as Gregg Easterbrook (1995) have questioned what he calls 'environmental orthodoxy' and have argued that while humanity faces environmental problems, we are hardly on the brink of global disaster.

Unfortunately, this emerging school of self-critical environmentalism has escaped the attention of many Christians focussing on environmental issues. In many respects, their writings underline the accuracy of Joseph Ratzinger's reflection that modern Christian intellectuals' interest in new social phenomena and secular movements can 'easily be deflected into the esoteric. It can evaporate in sheer Romanticism' (1988: 46).

Certainly, a relatively uncritical embrace of environmentalism has manifested itself in the thought of some Christian thinkers. Sean McDonagh, for example, has described what he believes to be the widespread destruction of bird life as an ecological and spiritual disaster. The churches, he maintains, should combat this by systematic moral teaching about 'biocide' (the elimination of species) as well as 'bird liturgies' that energise the human spirit (McDonagh 1990: 96).

One might charitably dismiss this as simply a case of extravagant rhetoric and somewhat dubious liturgical experimentation. But McDonagh begins to articulate quite questionable thoughts when he states:

> Gradually, it is beginning to dawn on many people that alleviating poverty, healing nature and preserving the stability of the biosphere is the central task for those who follow in the footsteps of Jesus in today's world. (1990: 163-4)

This proposition would seem to fall outside the bounds of Christian orthodoxy. Christians do have a responsibility to care for the poor and a responsibility to be careful in the way they treat nature. To claim, however, that healing nature and protecting the biosphere is the *central task* of Christians in the modern world is a highly suspect contention. The central task of Christians remains the same as it was in the beginning: to proclaim that Jesus Christ is Lord. For Christians, this announcement is of such earthly and transcendental significance that it cannot be reduced to a call for poverty alleviation and nature preservation.

Other Christian writers have absorbed the negative view of industrialisation and economic development articulated by some environmentalists. In an echo of Bahro's condemnation of industrialisation, the Protestant theologian Wesley Granberg-Michaelson states that he has 'come inevitably to the conclusion that the prevailing system is exploiting nature and peoples on a worldwide scale and . . . it is extremely urgent that we as churches make strong and permanent spiritual, moral and material commitments to the emergence of new models of society' (1992: 71).

The 'prevailing system' that Granberg-Michaelson has in mind soon becomes clear: 'environmental destruction and injustice', he alleges, 'have systemic causes such as the dominant development model itself with its emphasis on capital intensive industrialisation' (1992: 83).

Capitalism and the economic growth it facilitates thus stand directly accused. More generally, Granberg-Michaelson asserts: 'The whole notion of progress, economic growth and industrialisation, with escalating affluence, is the root of ecological destruction and the continuing impoverishment of millions' (1992: 14).

One could point out that an absence of economic growth is not likely to improve the poor's material well being. Moreover, Granberg-Michaelson seems unaware that the greatest environmental disasters of this century have resulted from collectivist rather than 'capitalist' designs. In the words of the Polish-Jewish sociologist, Zygmunt Bauman:

> Communism was modernity's most devout, vigorous, gallant champion . . . Indeed, it was under communist, not capitalist, auspices that the audacious dream of modernity, freed from obstacles by the merciless and seemingly omnipotent state, was pushed to its radical limits: grand designs, unlimited social engineering, huge and bulky technology, total transformation of nature. Deserts were irrigated (but they turned into salinated bogs); marshlands were drained (but they turned into deserts) . . . millions were lifted from "the idiocy of

> rural life" (but they got poisoned by the effluvia of rationally designed industry, if they did not perish first on the way). (Bauman 1992: 179)

The silence of many environmentalists about the scale of ecological destruction caused by the social engineering of command economies is deafening.

An even more radical position on environmental issues is taken by the former liberation theologian and now ex-priest Leonardo Boff. After Communism's collapse in 1989, Boff made a rapid transition from a heavy reliance upon Marxist analytical methods and hermeneutics to immersing himself in theories derived from deep ecology. There remains, however, a significant continuity in his work: the willingness to subordinate the Christian faith to the realisation of specific political agendas.

In *Cry of the Earth, Cry of the Poor* (1997) Boff maintains that we should be alarmed by population increases and the apparent decline of resource availability. He appears unaware that

> The truth is that the price of virtually every commodity—agricultural, mineral and energy—has fallen steadily throughout the twentieth century . . . A declining price is an indication of greater abundance, not greater scarcity. Food is so abundant today that the [American] government pays farmers not to grow so much. Of thirteen major metals, the only one that has risen in prices relative to wages in this century is platinum. . . . Fifty years ago the world had about twenty years worth of known reserves of oil. Thanks to technological innovation, which is outstripping the pace of depletion of reserves, the world now has at least fifty years of reserves. (Moore & Simon 1999: 29)

Apparently oblivious to these facts, Boff contends that declining resources and increasing population are dangerous because they threaten 'Gaia'. In Boff's view, 'the Earth is not a planet on which life exists . . . the Earth does not contain life. It *is* life, a living superorganism: Gaia' (1997: 24).

When it comes to articulating the theology that underlies this position, Boff verges on the pantheistic. If the Earth is Gaia with her 'force-fields' and 'morphogenetic fields', Boff claims, then God is 'that all attracting Magnet, that Moving Force animating all, that Passion producing all' (1997: 90). The language employed here clearly locates God *in* Gaia—the Earth as superorganism. This is contrary to the Judeo-Christian position that has always posited that the Divinity is *separate* from and *pre-existent* to his creation. A more nuanced but essentially similar position underlines the statement of Catharina Halkes, Emeritus Professor of Feminist Theology at the Catholic University of Nijmagen that, '[t]he image of the world as the body of God belongs more to our time and is closer to the changing reality than that of the Kingdom of God' (cited in Bandow 1993: 6).

Boff is, however, evidently aware that his ideas about the environment clash directly with basic Christian doctrine. This is apparent from his call for a revolution in Christianity's self-understanding:

> A revolution is successful only when it is the response to an urgent need for change; unless those changes are made, problems will continue, crises will deepen, and people will lose hope and meaning in their lives . . . a new spirituality, one adequate to the ecological revolution, is urgently needed . . . The conventional spirituality of the churches and of most historic religions is tied to models of life and interpretations of the world (worldviews) that no longer suit contemporary sensitivity. (Boff 1997: 139)

Significantly, Boff's demand for a dramatic change in Christianity's self-understanding is *precisely* the same message that he expounded in his past life as a liberation theologian. In *Church, Charism and Power*, for example, Boff stated that circumstances had superseded the Church's understanding of itself (he had in mind the Catholic Church). It now had to focus on making itself integral to 'the revolutionary situation' developing in the Third World by taking advantage of the 'revolutionary potential' found in the Church's memories of the 'subversive' Jesus Christ (Boff 1985: 87-92). The pope, bishops and priests, in Boff's vision of the Church, were no longer to be teachers but 'coordinators'. Boff proceeded to reject the image of the Church as mother and teacher as well as the idea of the Church as a sacrament of salvation on the grounds that such ideas suited the colonial world in which the Church identified itself with the status quo (Boff 1985: 152).

Reflecting upon these statements, the distinguished Jesuit commentator on Catholic social teaching, Rodger Charles, points out that Boff's version of liberation theology amounted to nothing less than 'a total rejection of [the Catholic Church's] self-understanding through the centuries down to and through the Second Vatican Council' (1998, vol. 2: 312). The common thread linking Boff's liberationist thinking with his 'eco-theology' is his belief that religion must serve the spirit of the times. As Boff himself puts it: religion 'cannot enclose religious persons in dogmas and cultural representations. It must serve as an organized place where people may be initiated, accompanied, and aided [in expressing] the spirit of the age' (1997: 214).

If, then, Christianity is to immerse itself in the apparently inevitable ecological revolution, Boff argues that Christians must strive to rid the world of any 'anthropocentricism'. This involves recognising that man is not *homo sapiens* (man the wise), but rather *homo demens* (man the deranged)—a creature whose demented state involves not realising that, in the wider schema of things, he is quite insignificant. In the forthcoming 'ecological and social democracy' for which the world is destined, Boff believes all religions must promote the notion that 'it is not just humans who are citizens but all beings . . . Democracy accordingly issues in a biogracy and cosmoscracy' (1997: 200).

Given Boff's disregard for the uniqueness of human individuals, it is hardly surprising that his response to environmental problems is to remove them from the domain of persons working for concrete solutions. Instead, Boff believes that all must be subordinated to 'global bodies, such as the United Nations and its eighteen specialized agencies and fourteen worldwide programs' (1997: 215). The principle of subsidiarity, it would seem, does not figure highly in Boff's scheme for reorganising the planet. Revealingly, the anti-private property/ anti-free enterprise motifs that manifested themselves in Boff's liberationist writings reappear in his conclusion as an 'eco-theologian', namely that only a new economic order based on the worldwide collectivisation of resources will save 'Gaia'.

While Boff's views about the environment take him far beyond the pale of orthodox Christian belief, less extreme but also questionable incursions of environmental thought into Christian theology are, as illustrated, not difficult to find. This makes all the more urgent the need for scholars to revisit the primary sources of Christian knowledge, such as Scripture, to discern what they tell Christians about how they should view and treat nature. In this regard, there is no more appropriate starting place than the book of Genesis.

In the beginning

At the root of environmental destruction?

Christian efforts to rethink aspects of theology and philosophy are often facilitated by challenges emerging from secular discourse. Much recent Christian reflection about the natural world's place in the Christian vision of the cosmos, for example, was prompted by a 1966 address by the cultural historian Lynn White to the American Association for the Advancement of Science. In this speech, White posited that the Judaeo-Christian tradition was largely responsible for contemporary ecological problems.

> Christianity, in absolute contrast to ancient paganism and Asia's religions, not only established a dualism of man and nature but also insisted that it is God's will that man exploit nature for his proper ends . . . Christianity made it possible to exploit nature in a mood of indifference to the feelings of natural objects. (1967: 1205)

White then asserted that people needed to 'find a new religion, or rethink our old one' if they wanted to undo the damage (1967: 1206). In his view, a starting point would be repudiating the apparently Christian view that humanity enjoys an 'absolute dominion' of nature that has no reason to exist except to serve man.

Similar explanations of Christianity's influence upon human attitudes towards nature have since been articulated at greater length. The work of John Passmore and Peter Singer features prominently in this regard. To an astonishing degree, Passmore's account (1974) of Western religious and ethical traditions as they apply to nature has been accepted as authoritative by many Western writers on environmental issues. The central theme of Passmore's writings is that the Bible, especially the Book of Genesis, teaches that people may use everything as they please. Christianity, it follows, encouraged a despotic view of nature, and led many to think that there are no moral constraints on man's dealings with the non-human world.

Passmore's interpretation lies at the heart of Singer's criticisms (though Singer invariably omits Passmore's careful qualifications). Singer acknowledges that many passages of the Hebrew Bible stress that people should be considerate of animals. Nevertheless, he maintains that the Genesis view goes unchallenged in Jewish and Christian teaching—a view which he portrays as holding that 'man is the pinnacle of creation [and] all the other creatures have been delivered into his hands' (1990: 205). Singer consequently insists elsewhere that

> According to the dominant Western tradition, the natural world exists for the benefit of human beings. God gave human beings dominion over the natural world, and God does not care how we use it. Human beings are the only morally important members of this world. Nature itself is of no intrinsic value, and the destruction of plants and animals cannot be sinful, unless by this destruction we harm human beings. (1991: 7)

Singer also believes that Genesis is incapable of encouraging Jews, Christians and Muslims to treat the environment carefully:

> Today Christians debate the meaning of this grant of 'dominion'; and those concerned about the environment claim that it should not be regarded as a licence to humanity to do as they will with other living things, but rather as a directive to look after them, on God's behalf, and be answerable to God for the way in which they are treated. There is, however, little justification in the text itself for such an interpretation; and given the example God set when he drowned almost every animal on earth in order to punish Noah for his wickedness, it is no wonder that people should think the flooding of a single river valley is nothing worth worrying about. (1991: 5-6)

It would be unkind to suggest that Singer displays in these statements a profoundly limited grasp of the Christian vision of the appropriate human attitude towards the natural world. Nonetheless, it is difficult to arrive at any other conclusion: the portrait presented by Singer (and to a lesser extent by Passmore and White) is at best a parody.

Unfortunately, it seems that few Christian—or, for that matter, secular—scholars have been willing to point out the extent of the inaccuracy of such interpretations, be they in the realm of history or biblical exegesis. This inactivity allows many myths to take hold of the academic, intellectual and popular imagination.

On a historical level, for example, the charges of Singer, White and Passmore against Christianity can barely be sustained. As Eric Jones explains, 'White offers no empirical verification that exploitative views are specific to Christianity; of the necessary corollary that religious thought of itself drives human actions; or that, *ceteris paribus*, exploitative actions have differed or differ among societies' (1991: 242).

Neither White, Singer nor Passmore acknowledges that ample evidence exists to demonstrate that ecological destruction has occurred in *non*-Christian cultures. Two of the most fertile (and non-Christian) areas in Asia—the plateau country of northern China and India's Ganges plain—are two of the most ecologically damaged parts of the world. In sharp contradiction to the myth that Taoist and Buddhist traditions protected the environment, Yi-fu Tuan's work details how the widespread destruction of forests and woodland 'tends to upset any residual illusion we may have of the Chinese farmer's benign attitude towards nature' (1968: 184). The pre-Christian peoples of Mesopotamia damaged much of the Fertile Crescent through the salination caused by their irrigation canals. Devotees of the Earth Mother ruined the landscape of Malta thousands of years before St Paul arrived on the island on his way to Rome (Malone et al. 1993: 76). Many pre-colonial peoples in America, such as the Maya, caused tremendous environmental destruction centuries before the Spanish conquest (Dubos 1984: 20-42; Elder 1996: 133).

Bernard Powell's study of Native Americans as conservationists illustrates that the stereotypical view of the Indians as mystics who did not pollute or litter the North American landscape is shown to be wrong by the litter at ancient campsites. Powell goes on to comment 'that all stereotypical views [of the Indians] are wrong, including specifically the view that Indians have instincts or culture norms as native ecologists, or insights into Nature denied to whites or other races' (1987: 17). Among other things, this is based on Powell's outline of extensive evidence of extinctions at the hands of the earliest Indians: buffalo dumps where the meat of one in every four bison filled was wasted; ruinous slash-and-burn agriculture by the Maya; the destruction of soils in the Desert Southwest by salination brought about by the irrigation farming of the Hohokam; the willing destruction of the beaver, the Indians' 'brother'; and the conspicuous waste of the Potlatch system in the Pacific Northwest.

One could go further and point out that some of the most 'anthropocentric' expressions of humanity's relationship to the animal and natural worlds are contained in *non-Christian* traditions. Stoics such as Cicero, for example, proclaimed: 'We are the absolute masters of what the earth produces' (*De Natura Deorum*, II: 60). Likewise, Confucius is recorded as stating that humanity's attitude towards the natural world is that 'we should control her course and use it' (*Analects*, trans. Leys 1997: 17,19).

It is also the case, however, that authors such as Passmore and Singer have an inadequate understanding of the Christian vision of the environment. This is especially true in Singer's case. His interpretation of the Noah narrative, for example, is simply incorrect. Any fair reading of the biblical story of the flood makes it clear that God did not punish Noah for his wickedness by drowning all the animals. The narrative records that God *saved* Noah *as well as* two of each of the animals *because* Noah, unlike the rest of humanity (who were *also* drowned alongside almost every form of animal and plant life) was a *just* man—'Noah was a good man, a man of integrity among his contemporaries, and he walked with God' (Gen 6:9-10).

Moreover, when one turns to examining texts that are the target of Singer's ire, it soon becomes apparent that his interpretation does not stand up to critical analysis. Instead, Genesis outlines a basis for viewing the world that both encourages human creativity and economic development as well as a responsible attitude towards the environment.

The Hebrew de-divinisation of nature

Before engaging in closer analysis of Genesis, we should note just how different the vision of God, humanity and nature expressed in this part of Scripture was from the dominant view of the cosmos prevailing in the ancient world. Singer is correct when he states that Genesis places man at the pinnacle of creation. It does so precisely by stripping nature of the divine status that it was usually accorded in the pre-Christian world, and investing the human person with a unique dignity. Passmore arrives at a similar conclusion: 'The view that man in any sense rules over nature inevitability presumes that nature itself is not divine. And the striking peculiarity of the religion of the Hebrews, when we compare it with the Middle Eastern religions which surrounded it, is its distinction between God and nature' (1974: 10).

Though Passmore is not a Scripture scholar, his analysis accords with that of the eminent biblical exegete, Gerhard von Rad. According to Rad: 'Investigation of the cultures and religions of Israel's neighbours shows that she is absolutely unique in taking man out of the sphere of myth. She dropped the mythological realm of spirits and magical powers . . . Even Israel's kings were not a mythic primeval datum of the created order as was the case with some of her neighbouring nations' (Rad 1960: 349). Jewish scholars such as the Chief Rabbi of the United Kingdom and the Commonwealth, Professor Jonathon Sacks, take a similar view:

> It was Max Weber who observed that one of the revolutions of biblical thought was to demythologise, or disenchant, nature. For the first time, people could see the condition of the world not as something given, sacrosanct and wrapped in mystery, but as something that could be rationally understood and improved upon. (1998: 16)

This de-mythologisation of nature accompanied by a particular emphasis upon human dignity was surely a welcome development, given the widespread contempt for human life that dominated the pre-Christian world (Johnson 1976: 12; Gregg 1999b: 7), be it in the Americas, the Mediterranean or the Middle East. The Israeli writer Hannes Stein points out, for example, that

> The Aztecs were not exceptionally cruel and only a fool would call them uncivilized or barbarian. But human sacrifice was precisely what defined the ancient civilizations. Thus the Canaanites threw children into fiery furnaces to please Moloch: the Egyptians worshipped the sun and the goddess Hathor . . . the Assyrians and Babylonians built the first cities around enormous slaughterhouses where priests sang praise to the stars before they cut the throats of well-built young men. How could the Israelites with their nomadic ancestor Abraham compete with this? The Philistines, by comparison a civilized race, prostrated themselves before their fish-god Dagon. They were immigrants from Crete, where the celestial bull demanded the lives of a dozen virgins a year.

> Human sacrifices were no cause for shame. They were not performed discreetly in a clandestine cellar but on top of a pyramid, in the temple, in front of a crowd. Lo and behold, we are prepared to give what is most dear to us! Look, we do not even spare our children. So voracious were the star gods. So great was the fear of the pagans. And it could have gone on forever according to the eternal cycle of nature, accompanied by the howl of shamans, the singsong of vestal virgins, and the roar of the slaughtered. (1999: 35)

The various Canaanites cults that surrounded the Hebrew people worshipped the fertility goddess Ashtaroth and her Baal consorts by paying homage to many places, symbols and animals which, in their view, expressed the numerous revelations of the nature deities (Rad 1957: 227-8). Against this, Hebrew texts such as Deuteronomy (12: 2-7) emphasised that Yahweh was *one* and far greater than graven images of bulls and other creatures. Hence, as Ratzinger states:

> The faith of Israel is certainly something new in comparison with the faith of the surrounding peoples . . . Yahweh, their God, is an only God—this fundamental confession . . . is in its original sense a renunciation of the surrounding gods. As a renunciation of the gods it also implies the renunciation both of the deification of political powers and of the deification of the cosmic . . . a renunciation of the fear that tries to tame the mysterious by worshipping it . . . (1968: 73-4)

Judaism's emphasis upon God's transcendence over the world was transmitted to Christianity. It is reflected in the Apostles Creed in which God is described as 'Almighty'. The phrase is derived from the Greek *pantokrator* and the Hebrew Bible's *Yahweh Zebaoth*. Literally translated, it means something like 'God of hosts' or 'God of powers'. For our purposes, its significance lies in the fact that, as Ratzinger comments, 'For all the uncertainties about its origin we can at any rate see that this word is intended to describe God as Lord of heaven and earth; it was probably intended above all to define him, in opposition to the Babylonian religion of the stars, as the Lord to whom the stars belong, alongside whom the stars cannot exist as independent divine powers: the stars are not gods, but *his* tools' (1968: 103).

This vision of the world was given a particularly polemical form in the first and second commandments of the Decalogue:

> You shall have no gods except me. You shall not make yourself a carved image or any likeness of anything in heaven or on earth beneath or in the waters under the earth; you shall not bow down to them or serve them. (Ex 20:3-5)

According to Rad, these ancient words made it inevitable that Israel would challenge 'the idea that the world is a place where there were a number of ways by means of which God directly revealed himself. She knew that she was poles apart from the basic presupposition of all forms of idolatry, namely, the belief that the divine nature is embodied in a variety of earthly embodiments and cultic symbols' (1960: 339).

From this standpoint, the first and second commandments represented a declaration of war on the nature gods. In Israel's eyes, the world was *not* constructed as a realm of fixed quasi-sacral orders that rested on the shoulders of a multiplicity of quasi-divine powers. Ancient man's seemingly endless capacity to objectify his experiences of the elemental powers of the natural world in which he lived and to regard these as divine, was fiercely resisted in Israel. Indeed, having stripped the natural world of its semi-divine status, the Hebrew Scriptures underline Yahweh's *authority* over the world: He looks upon the earth and it trembles. He touches the mountains and they smoke (Ps 104:32). He shakes the earth, so that its pillars tremble, and forbids the rising of the sun (Job 9:4).

Moreover, if, as Scripture maintains, the world is the product of God's creative word, then it is sharply separated in its nature from God himself. To locate God in the Earth, as Boff does, is to ignore what the three great monotheistic faiths regard as the sheer awesomeness of God's creative Act. As St Theophilus of Antioch wrote:

> If God had drawn the world from pre-existent matter, what would be so extraordinary in that? A human artisan makes from a given material whatever he wants, while God shows his power by starting from nothing to make all he wants. (*Ad Autolycum*, II:4 [in *Patrilogia Graeca*, 6:1052])

The precise place of nature

The Hebrew de-divinisation of the world of nature was not, however, a mandate for wanton ecological destruction. Elder points out that the Israelites had a rather good ecological record (Elder 1996: 132). In biblical times, for example, they let the land lie fallow and open to wildlife every seventh year' (Ex 23: 10-11). Numerous Old Testament passages such as Ps 104 and Job 39-40 repudiate a despotic view of nature on the part of humans.

What does Scripture tell us about nature's place in the Judaeo-Christian vision? Scrutiny of Genesis soon indicates that animism is not, as White held, the only form of religion which has prohibited excessive exploitation of nature (White 1967: 1205). As the Anglican theologian and economist, Lord Griffiths notes, the two crucial facts about the natural world from Genesis's standpoint are that it is God's and that it is intrinsically good (Griffith 1984: 50). Nevertheless, while Genesis considers nature to have a value independent of man, it is a value dependent on God and not superior or equal to man. In summarising this position, Rad comments:

> The various works of Creation stand on a completely different footing in respect of their relationship to the Creator—they are far from having a like immediacy to God. At farthest remove from him, in a relationship which scarcely admits of theological definition, is the formless, watery, darksome, abysmal chaos . . . The

> plants have a very indirect relationship to God, for they spring from the ground, which God commissioned to play a part in creating them. The animals also have an immediate relationship to the ground, but they are the recipients of a special word of blessing assigning fruitfulness to them, in order that they may multiply. (1957: 142-3)

In making the heavens and the earth, God declared the prehuman creation to be good without anticipatory reference to man: 'And God saw that it was good' (Gen. 1: 10, 12, 18, 21). The same point is made elsewhere in Scripture (Ps 19, 33, 104:24, 148; Prv 8: 25-31 Wis 11:24-25). It is therefore consistent with Jewish, Christian and Islamic tradition that the earth, sky and all living things have a certain value of their own, although they do not possess the distinctive value conferred upon humanity by virtue of their dignity as the *imago Dei*. Each of the various creatures, willed in its own being, reflects in its own way a ray of God's wisdom. Humans must therefore respect the particular goodness of every creature and avoid any disordered use of things, for this would be to show contempt for the Creator.

Humanity and nature

If one accepts that nature has a value of its own, what is the appropriate human attitude towards it? Paul Collins, M.S.C., claims:

> Christians have always believed that the natural world is God's creation and that its splendour and complexity mirrors God's splendour, and that to destroy that world, for whatever reason, is to destroy one of our most precious images of God. (1999/2000: 9)

Collins' statement is correct insofar as it underlines the fact that the beauty of creation reflects the Creator's infinite beauty. Nature ought therefore to inspire the respect of human beings. 'For from the greatness and beauty of created things comes a corresponding perception of their Creator' (Wis 13:4).

Yet as Ian Hore-Lacey notes, acknowledgment of the world's beauty 'does not mean that preservation of natural areas from human influence is *prima facie* their highest use' (1985: 48). Moreover, Collins' statement underestimates, if not ignores, the beauty forged by man. The original wilderness of Europe, for example, has been transformed. But would anyone seriously question that so many of the ensuing human creations speak of a timeless beauty and inspiration? In short, it should be recognised that when they enter the world, people enter into two inheritances: the inheritance of what has been given to everyone in the resources of nature, as well as the inheritance of what others have already developed on the basis of such resources. The aesthetic value of the latter should not be dismissed so quickly.

Moreover, as any serious Scripture scholar knows, the *only* creature specified in Genesis as made in God's image is *man* (Gen 1: 26-27; 2: 5-8). To accord this dignity to other creatures is therefore dubious, not least because it tends to undermine the Christian belief—so clearly set out in Genesis—that humans do indeed enjoy dominion over the earth.

The dominion, however, that Christians believe that humans enjoy over the earth is not quite the dominion portrayed by Singer, Passmore and White. As noted by the Scripture scholar Thomas Dailey, O.S.F.S., the dominion given to humanity is not an unrestricted power (Dailey 1992: 1-13).

There are two accounts of man's creation at the beginning of the book of Genesis. The first occurs in chapter one and is thought to have been ordered or edited by the priests of the Jewish people in exile in Babylon (Schmitz 1993: 93-4). Hence, it is called the 'Priestly' account. This situates man's creation within the account of the world's creation in seven days, with man being created on the sixth day (Gen 1: 26-29). The second account (in the second and third chapter of Genesis) is considered to have been written earlier, possibly in the eighth century B.C. Because it uses the name Yahweh to refer to God, it is known as the 'Yahwist account' of Creation (Gen. 2: 5-25).

Both accounts specify that man is the summit of the Creator's work, as both distinguish the creation of man, male and female, from that of the other creatures (Gen. 1: 26-31; 2: 6-7), with the result that the rest of the world is ordered around humanity as Yahweh's chief work. This hierarchy of creatures is expressed by the order of the 'six days' from the less perfect to the more perfect. But while God loves all his creatures (Ps 145: 9), Rad points out that

> At the top of this pyramid stands man, and there is nothing between him and God; indeed, the world, which was in fact made for him, has in him alone its most absolute immediacy to God . . . God was actuated by a unique solemn resolve in the depths of his heart. And in particular, God took the pattern for this, his last work of Creation, from the heavenly world above. In no other work of Creation is everything referred so very immediately to God himself as in this. (1957: 142-3)

This vision of humanity's uniqueness is underlined in the New Testament. Not only does it portray God as *becoming* man, but it is full of statements such as 'You are of more value than many sparrows' (Luke 12: 6-7), and notes 'how much more value is a man than a sheep' (Mt 12: 12).

There are, nevertheless, differences in nuance between the Priestly and Yahwist accounts which, taken together, provide a fuller understanding of the Genesis view of man's relationship to the natural world. The key words in the Priestly account read:

> God blessed them, saying to them, 'Be fruitful, multiply, fill the earth and subdue it [*kabaš*: to tread down, subjugate], and rule over it [*radâ*: to tread down, have dominion]. Be masters of the fish of the sea, the birds of heaven and all living animals on earth'. (Gen 1: 28)

These are strong words. They leave us in no doubt as to man's authority over the created world. Reflecting on this verse,

Karl Barth—regarded by many as the greatest theologian of the twentieth century—states:

> God set man in the world as the sign of his own sovereign authority—it was in that sense that Israel thought of man as the representative of God. . . . This lordship of man extends over the world and not, for example, just over animals. The reason why the animals are mentioned is because they alone come into question as the rivals of man. But they are expressly put under him. (1961: 206)

From the standpoint of a different Christian tradition, Germain Grisez makes a similar point when he writes that '[humans] are responsible for [nature], but not to it, as if it shared in the dignity and fundamental rights which they themselves enjoy as persons made in God's image' (1993: 775).

Turning to the Yahwist account, we observe that while it does not question human dominion over the earth, it does provide some parameters as to *how* this control should be exercised:

> Yahweh God took the man and put him in the Garden of Eden, to work it ['*abad*: to serve, till, enslave] and to keep it [*šamar*: to hedge about, protect, guard]. (Gen 2: 15)

Here we observe the Priestly account's emphasis upon human authority being integrated with a concern for good management and preservation. What the Genesis narratives do *not* tell us is precisely what particular parts of the natural world are best suited to man's service and in what ways these may be actualised. Nor do they inform us which parts of the world are best left preserved. It consequently seems that God gave humans a certain freedom concerning the ways in which they use the earth, not least because through use of their intelligence and free will, humans can help complete the work of creation. For while creation has its own goodness and proper perfection, it did not, as the *Catechism of the Catholic Church* notes, 'spring forth complete from the hands of the Creator. The universe was created in a state of journeying (*in statu viae*) toward an ultimate perfection yet to be attained, to which God has destined it' (Catechism 1994: para. 302).

There are, moreover, clear implications scattered throughout Scripture that God's people are supposed to do something with the natural world. The Hebrews' wandering in the desert (Deut. 8: 2-24), for example, occurs in the context of a Promised Land waiting for them in which they will live and work. Work, of course, usually involves more or less directly interacting with and transforming sub-personal things. It manifests humanity's participation in God's superiority over nature. Genesis's attention to work underlines that the Jewish and Christian traditions do not believe that man was created immobile and static. The Bible's first portrait of man presents humans as creatures whose uniqueness as the *imago Dei* is underlined in part by the fact that they alone can work.

The command to work

In the Greek world, material work was regarded as a necessity not fit for free men, who engaged in politics and philosophy. Only slaves worked (Charles 1982: 312-3; Gregg & Preece 1999: 17). Judaism and then Christianity changed this view of work in the West forever. From the beginning, Jewish theology held that people are called upon to work in order to fulfil themselves and make known God's majesty throughout the universe. The Jewish liturgy for Saturday night—the point at which the Sabbath day of rest ends—culminates in a hymn to the value of work: 'When you eat of the labour of your hands, you are happy and it shall be well with you'. The Jewish tradition also maintains that work has spiritual value because, as Rabbi Sacks notes, 'earning our food is part of the essential dignity of the human condition. Animals *find* sustenance; only mankind *creates* it' (1998: 15).

The very first pages of Genesis indicate that work is a fundamental dimension of human existence on earth. In Barth's words, work is 'the distinctly this-worldly element in the active life required of man' (1961: 2). Genesis specifies in unambiguous language that people are not made for a life of leisure: they are *commanded* to work. It is here that Christianity's distinctly anti-Rousseauian dimension becomes manifest. Made in the Creator's image, people are charged with the responsibility of unfolding the Creator's work. Man is told to 'Be fruitful and multiply, and fill the earth and subdue it' (Gen 1: 28). These verses indirectly underline work as an activity for people to carry out in the world.

Humans are thus mandated to control and harness the forces of nature. This may range from bringing wasteland into cultivation and improving the productivity of existing farmland, to extracting minerals and using them in a manufacturing process. Such processes may be abused—monopoly, corruption, fraud, pollution—and people should be held accountable for this. Nevertheless, the legitimacy of the process is not judged by its abuse: we do not condemn, for example, eating because of gluttony, sex because of adultery, or property because of greed.

Animal 'rights'?

Genesis's vision of humanity's relationship with the environment is evidently more nuanced than some secular and Christian commentators would have us believe. What, however, are the implications of such a view to particular issues often associated with modern environmental activism?

The animal 'rights' agenda is one that has been embraced by sections of the Green movement. In 1969, for example, Paul Ehrlich—infamous for making apocalyptic predictions about impending doom that never occur—predicted that all major animal life in the sea would be extinct by 1980 (Elder 1996: 128). More importantly, Ehrlich and others have employed various forms of moral vocabulary to articulate what they as environmentalists believe to constitute the most appropriate moral relationship between humans and animals. Ehrlich, for example, has stated that 'people have an absolute moral responsibility to protect our only known living companions in the universe' (Ehrlich & Wilson 1991: 761).

The primary problem with this particular statement is its use of the word 'absolute'. Adherence to such a position would mean that humans were not allowed to use animals in any way, a view evidently at odds with the Judaeo-Christian position outlined in Genesis.

There are, however, some who believe that Christianity essentially encourages exploitative attitudes towards animals. This form of criticism is exemplified by Singer's claim that 'the New Testament is completely lacking in any injunction against cruelty to animals on any ground, or any recommendation to consider their interests' (1990: 209).[3] He also argues that, after Plutarch, 'We have to wait nearly sixteen hundred years . . . before any Christian writer attacks cruelty to animals on any ground other than it may encourage a tendency toward cruelty to humans' (1990: 211).

We have seen, however, that Genesis clearly suggests that animals have a value and that they are not to be misused or abused. Moreover, contrary to Singer's statements, similar ideas *are* affirmed in the New Testament (John 10: 11; Romans 19-20; Col. 1: 15-20; Rev 5: 13). This care for animals is also expressed in Christian texts ranging from the prayers of St Francis of Assisi to Eastern Christian scholars of the first millennium. Even Passmore acknowledges this when he cites St Basil's prayer for animals:

> And for these also, O Lord,
> The humble beasts, who bear with us the heat and burden of the day,
> We beg Thee to extend Thy kindness of heart, for Thou has promised to save both man and beast,
> And great is thy loving-kindness, O Master. (Passmore 1975: 198)

This prayer supplies a reason for kindness to animals that is entirely independent of human interests. So too does the remark of the fifth century saint, St John Chrysostom, that 'we ought to

[3] For a devastating critique of Singer's historical understanding of Christianity's view of animals by a secular thinker, see Attfield (1983: 202-10).

show them [animals] great kindness and gentleness for many reasons, because they are of the same origin as ourselves' (Linzey 1976: 103). Basil and Chrysostom's advocacy of compassion for animals is replicated in the seventh century teaching of St Isaac the Syrian (Allchin 1978: 85), not to mention St Thomas Aquinas' condemnation of the misuse of animals (*Summa Contra Gentiles*, Pera 1961: 3.112).

More recently, many churches have stated at length that Genesis does not provide a mandate to abuse animals. According to the *Catechism of the Catholic Church*, 'Animals are God's creatures. He surrounds them with his providential care. By their mere existence they bless him and give him glory (Mt 6: 26). Thus men owe them kindness' (Catechism 1994: para. 2416).

At the same time, however, the *Catechism* insists:

> God entrusted animals to the stewardship of those whom he created in his own image. Hence it is legitimate to use animals for food and clothing. They may be domesticated to help man in his work and leisure. Medical and scientific experimentation on animals, if it remains within reasonable limits, is a morally acceptable practice since it contributes to caring for or saving human lives. (Catechism 1994: para. 2417)

The welfare of animals, then, yields before the well being of humans, and prudence is the key to discerning when legitimate use degenerates into misuse.

In Singer's view, however, any such argumentation does not merit consideration. Its theistic basis is simply 'irrational' and therefore automatically excluded from any reasonable discussion. This, however, is yet another example of what one commentator describes as

> . . . Singer's bad habit of treating as a settled matter issues over which reasonable people disagree. We know, Singer explains, as if he were reciting the obvious for the umpteenth time, that men and women were not created in God's image, because . . . well, as a matter of fact Singer never does say how it is that we do know

> for sure, how rationally and in good faith we can conclude once and for all that the truths of faith are altogether empty and untrue. (Berkowitz 2000: 33)

Having asserted that the atheistic conclusion about the origin of the universe is (apparently) self-evidently true, Singer maintains:

> If the universe has not been constructed in accordance with any plan, it has no meaning to be discovered. There is no value inherent in it, independently of the existence of sentient beings who prefer some states of affairs to others. Ethics is no part of the structure of the universe in the way that atoms are. (Singer 1993: 188)

This being the case, Singer's view of animals is based on the (again, apparently self-evident) principle that a coherent system of ethics must be grounded in the principle of equal assessment of interests. Hence, just as we treat intelligence differences in people as morally irrelevant, so too should we consider features such as four-leggedness and the inability to speak as having no place in moral discourse. The only morally relevant factors for Singer are the capacity for pain and pleasure as well as sentience (again, no reason is given as to *why* these are so important). As animals possess all of these, Singer insists that they are entitled to have their interests considered within our moral order.

In general terms, Singer's argument is that people have rights because they have interests which others' actions can fulfil (leading to pleasure) or frustrate (leading to pain). But, Singer suggests, animals to varying degrees—higher animals more (e.g. a horse), lower animals less (e.g. a snail)—also have interests whose fulfilment or frustration causes them pleasure or pain. On this basis, Singer asserts that animals have rights (1990: 6-9, 17-20).

To suggest anything to the contrary, in Singer's view, 'is to give preference to the interests of members of one's own species, simply because they are members of one's own species. This is speciesism, a moral failing that is parallel to racism, because it attempts to put a morally crucial divide in a place that is not justified on any basis other than a preference for "us" over "them"' (1991: 15). The logic of this position allows Singer to maintain that while any mature and normal animal has some 'rights', unborn and newborn human beings have none whatsoever (1990: 81-2, 236-43).

The idea that animals have rights precedes the theories of Peter Singer. It was promoted, for example, by one of the earlier and most prominent heresies, the Manichaeans, for which they were rigorously condemned by St Augustine (*De civitate Dei* 1: 20). The Christian view of rights proceeds from a different basis and arrives at different conclusions. Broadly speaking, it holds that humans are different from subpersonal creation because they are made in God's image. This is reflected in a nature that, among other things, includes the unique capacity for reason and free choice. 'As such', according to John XXIII, 'man has rights and duties, which together flow as a direct consequence from his nature. These duties and rights are universal and inviolable, and therefore inalienable' (1963 [1981]: para. 9). Christian ethics does not therefore agree that fundamental human duties and rights are derived from sensory awareness or other features that humans have in common with animals.

Singer, of course, rejects the position that humans are endowed by God—whom he simply presumes that no reasonable person could believe to exist—with any fundamental rights. He also minimises the importance of the difference between human reason and animal cognition, and insists that any distinction between human interests and those of other creatures only provides a basis for specifying which rights can be possessed by various individuals (1990: 187-98).

But one does not have to accept the Christian view of the moral life to grasp some of the foundational problems that bedevil Singer's contentions. Any coherent theory of duties and rights—secular or religious—presupposes creatures capable of defining and respecting morality. Animals, however, are by nature incapable of this. They cannot even construct a morality out of their experience of pain or pleasure. Hence, given that animals cannot know, respect, or exercise duties and rights (either in actuality or potentiality), the idea of duties and rights is inapplicable to animals.

One could add here that if animals had rights, humans would have corresponding duties. But when asked to yield their interests to those of animals, one would expect most humans to respond negatively. 'Why', many would ask, 'should I be prevented from using my land because of its current occupation by a particular species of animal?' The challenge, then, for animal rights advocates is to provide an account of *moral obligation* adequate to show why—given *their* understanding of rights—any moral agent ought to respect anyone else's rights.

As Grisez notes, it is at this point that the animal rights argument begins to disintegrate because animal rights advocates *cannot* provide an account of moral obligation:

> [o]n their view, each agent naturally acts egotistically in accord with his or her interests, while naturally serving that of others only insofar as they coincide with or are embraced in his or her own as, for example, the interests of friends and family often are.
>
> In this view, however, moral obligation is the demand that agents act altruistically when such a natural motivation is lacking and even when doing so is contrary to their interests. But in that case, why should anyone be moral? Psychological and sociological attempts to account for moral feelings and practices do not begin to answer this question. At best they can explain only why some people in fact feel or think they ought to be altruistic. But the question is: Why *should* egoists repent and become altruists.
>
> No thinker sharing the general worldview of the proponents of animal rights ever has offered a plausible answer. For them, moral obligation remains inexplicable. (Grisez 1993: 784)

None of this means that humans are free to misuse animals. It does, however, mean that Christians and others can dispute proponents of animal 'rights' on their own terms, and be confident that, in principle, it is permitted to use animals for human food, clothing, shelter etc, provided that

- the animals are used for human benefit to the extent necessary for the purpose in mind or the use is unavoidable without imposing significant burdens on humans; and
- the act is not morally wrong on other grounds.

Thus, while it may be morally questionable for a person to wear a fur coat simply for ostentatious display, the use of animal fur to keep warm is a different matter. Certainly, our treatment of animals cannot be indiscriminate and unrestricted because it represents misuse of the gift given to man. But it does not mean that we are committed to 'giving' rights to animals. One could even posit that ascribing moral status to animals is precisely an example of 'speciesism' insofar as it amounts to an imposition of human moral discourse upon animals.

Some challenges to environmentalism

Since the emergence of environmentalism as a political and philosophical force, the Christian churches have expended much time responding to environmentalist critiques of the orthodox Christian view of the cosmos. One may contend, however, that many have failed to recognise that there are several areas in which Christians could quite legitimately be particularly *critical of* the green lobby. Two matters feature prominently. One concerns the attitude of many environmentalists towards population issues; the second involves their inadequate grasp of what might be called the extent of the natural order—more specifically, the apparent disinterest of some environmentalists in sustaining what might be called humanity's *moral* ecology.

A population 'problem'?

One theme commonly underlined in many environmentalist writings is the need to stabilise, if not halt and reverse, population growth. In many instances, this is based upon the presumption that the more humans that exist, the more the environment will suffer. In doing so, some environmentalists tend to endorse dramatic predictions about the earth's future, such as the claim of four MIT academics that

> If the present growth trends in world population, industrialization, pollution, food production, and resource depletion continue unchanged, the limits to growth on this planet will be reached sometime within the next 100 years. The most probable result will be a sudden and incontrollable decline in both population and industrial capacity. (Meadows et al. 1974: ix-x)

The Global 2000 report made similar predictions about the world's population and natural resources. In every resource category, Global 2000 predicted overuse and declines in quantity and quality (1980).

Fundamentally, such claims are premised upon the assumptions made by the eighteenth century economist Thomas Malthus: that human population growth is liable to outstrip the means of subsistence ([1798] 1992). The tone, however, of some environmentalist writings on this matter verges at times on the apocalyptic (Efron 1984). Prominent environmentalist Al Gore has, for example, referred to population growth as precipitating an 'ecological Kristallnacht' (1993: 110)—an emotivist comment that many would regard as cheapening Kristallnacht's significance.

Such predictions have led some environmentalists to reflect upon how the world might cope with the apparently inevitable problems proceeding from population growth. Some have indicated that their preference is simply to have fewer people around. J. Baird Callicott, for example, claims that '[i]f it is not only morally permissible, but, from the point of view of the land ethic, morally required, that members of certain species be abandoned to perdition . . . or even culled, how can we consistently exempt ourselves from a similar draconian regime? We too are only "plain members" of the biotic community' (1989: 92). More chilling are statements by animal rights activist Tom Regan. He insists that '[m]assive human diebacks would be good. It is our duty to cause them. It is our species' duty, relative to the whole, to eliminate 90% of our numbers' (Regan 1990: 296). Similarly, the biologist David Graber is quoted as suggesting that humans 'have become a plague upon ourselves and upon the earth. . . . Until such time as *homo sapiens* should decide to rejoin nature, some of us can only hope for the right virus to come along (cited in Postrel 1990: 28).

Though not quite advocating such solutions, concerns about the effect of growing population upon the environment have manifested themselves in Christian theological discourse. The feminist theologian Anna Primavesi, for example, claims that 'human excesses are now coming together exponentially, in a catastrophic relationship between human fertility and the earth's humanly imposed infertility' (1991: 13). She does not, however, provide any statistical evidence to demonstrate that this is indeed the case. Instead, Primavesi emphasises that 'Christian values, with their destructive lack of ecological wisdom, are no longer perceived by other systems of thought as having any positive role to play in the present world crisis' (1991: 14).

As we have seen, Christianity does in fact contain much 'ecological wisdom' of which Primavesi is apparently unaware. More importantly, there is no reason why churches should allow themselves to be stampeded by environmental Malthusians into calling for immediate and drastic population controls. Thomas Sowell, for example, has demonstrated that there is very little statistical correlation or causal relation between poverty and high and dense population. Certainly, he states, India is poor and heavily populated while Kuwait is rich and sparsely populated. Yet there are numerous opposite examples. Millions live in London and New York, while many underpopulated countries such as Somalia live in poverty. In 1981, Hong Kong with its free economy comfortably supported 14,000 people per square mile, while then-socialist Ethiopia struggled to keep 100 people per mile alive. (Sowell 1983: 208-217). In short, Julian Simon appears to be correct when stating that

> Population growth does not have a statistically negative effect upon economic growth. We know that from 30 years of careful quantitative scientific studies—just the opposite of what the public believes. Because human knowledge allows us to produce more finished products out of fewer raw materials, natural resources are

> becoming more available. The air and water in rich countries is becoming cleaner. Most importantly, human beings are living much longer than ever before. (1995: 1)

The perennial problem of Malthusian arguments is that they rely heavily upon the premise that demands on resources will be endless while supply is finite. Such forecasts fail to take account of man's ability to react to problems of scarcity by reducing consumption, finding substitutes and improving productivity. The human mind is, in Simon's words, 'the ultimate resource', which has permitted us to avoid the Malthusian trap (Simon 1996). This view is echoed in John Paul II's social encyclical *Centesimus Annus* which reminds us that 'besides the earth, man's principal resource is man himself' (John Paul II 1991: para.32). Indeed, Charles reflects that history demonstrates that

> Human ingenuity has never failed yet in developing the world and its wealth for its purposes or finding substitutes for things in short supply, and there are no rational grounds for thinking it will not do so in the future; indeed with the increase of knowledge and technology at our command, our power to do this is greater than ever. (Charles 1998, 2:150)

In recent years, concerns have been raised by environmentalists themselves about the inaccuracy of many predictions of the Green lobby. Easterbrook, for example, notes that the overwhelming majority of forests in Europe and America have *not* been destroyed by pollution. Fossil fuels have *not* been exhausted. Growing populations have *not* caused worldwide food shortages. Nor have wildlife species been made extinct on a massive scale (Easterbrook 1995; Budiansky 1996). In 1972, for example, the Club of Rome asserted that humanity's existence was threatened because of the imminent depletion of resources. Yet not only does the empirical evidence illustrate that *all* the significant resources that the Club of Rome identified as eventually running out have actually increased, but the Club of Rome itself eventually disowned its 1972 statements (Simon

& Kahn 1984: 104). There is now more known stocks of oil, natural gas, coal, and water in the world than there was twenty years ago (Hodel 1997: 1-4). Simon agrees:

> Every agricultural economist knows that people have been eating better since World War II in the period for which we have data. Every resource economist knows that natural resources have become cheaper rather than more expensive. Every demographer knows that life expectancy in the wealthy countries has gone up from under 30 years at birth 200 years ago to over 75 years at birth today. And life expectancy has risen in the poor countries from perhaps 35 years at birth only 50 years ago to 60-65-70 years at birth today. (1995: 1)

If we examine countries which have been badly affected by famine and other natural disasters in the last twenty or thirty years, it soon becomes evident that it is the absence of good government, peace and stability which causes starvation—not 'overpopulation'. To this, one could add over-militarisation, 'over-regulation, over-bureaucratisation and over-politicization' (Block 1999: 281), social engineering experiments, and socialist economic policies. To illustrate this point, Syen Rydenfelt analysed the agricultural performance of 16 socialist states including Cuba, Tanzania and China over a twenty-year period. In each case, the result was the same: massive downturns in food production. In ten years, Julius Nyerere of Tanzania managed to reduce a nation once self-sufficient in food and actually exporting corn, to dependence upon foreign aid in food stuffs (Rydenfelt 1984). In short, it is not overpopulation that produces famine but antiquated political and economic systems.

A similar picture emerges from analyses of world food prospects. Though Ehrlich (1968) predicted that mass starvation would ensue within a decade of the publication of his book (and, even more predictably, urged strict population controls as a solution), it is apparent that (once again) Ehrlich was wrong. Moreover, as Ron Duncan notes, '[c]oncerns about an impending or even distant global imbalance between population growth and food supply are exaggerated. The major problems in the food supply system are either man-made or can be corrected through institutional development' (1998: 80). Duncan illustrates that food supply problems have less to do with 'overpopulation' and much more to do with poor policies and the absence of effective property rights.

What, then, should be the Christian role in light of such facts? For one thing, it may be suggested that the churches should *inform themselves* of these details so that they do not make unsubstantiated observations similar to those of Primavesi. More significantly, Christians, along with Jews and Muslims, can underline in any discussion about population one of the many fundamental insights contained in Genesis: that each human being is not simply a consumer but a potential *creator*—a being that innovates, thinks and freely acts. Instead of regarding people as dangerous to the environment, Christians can stress that we need to stop viewing people as a burden but rather as an *asset*. Here, Christians should be prepared to bring to public attention the gross violations of human dignity that have accompanied many attempts at population control. These include state-enforced one-child policies, not to mention episodes such as India's sterilisation programme of the 1970s (Johnson 1983: 570-1).

The priority of human ecology

The second criticism of much contemporary environmentalism that Christians could underline is the (paradoxically enough) highly materialistic view of the natural order that informs much environmentalist thinking. Environmentalists often ignore what Robert George of Princeton University and Michael Novak have called the *moral* ecology that underpins free societies (George 1993: 42-47; Novak 1999: 1-6).

As noted, environmentalism's emergence may owe much to humanity's unease with how the world has developed. An expression of this unease is the insistence of many—within and outside environmental movements—that humans are not entitled to do whatever they will with the world of nature. Figures such as John Paul II have stated, however, that the same point can be made with regard to humanity's moral environment:

> In addition to the irrational destruction of the natural environment, we must also mention the more serious destruction of the human environment, something which is by no means receiving the attention it deserves. Although people are rightly worried . . . about preserving the natural habitats of the various animal species threatened with extinction . . . too little effort is made to safeguard the moral conditions for an authentic 'human ecology'. Not only has God given the earth to man, who must use it with respect for the original good purpose for which it was given to him, but man too is God's gift to man. He must therefore respect the natural and moral structure with which he has been endowed. (John Paul II 1991: para 38)

In 1996, Ratzinger applied the principles underlying this argument—i.e. that there is a natural human moral ecology which contains just as many unbreakable laws as the material world—to highlight what he views as a major omission in the thinking and pronouncements of many contemporary environmentalists:

> I think that this is . . . the defect of the ecological movements. They crusade with an understandable and also legitimate passion against the pollution of the environment, whereas man's self-pollution of his soul continues to be treated as one of the rights of his freedom. There is a discrepancy here. We want to eliminate the measurable pollution, but we don't consider the pollution of man's soul and his creaturely form. Instead of making it possible to breathe humanly again, we defend with a totally false conception of freedom everything that man's arbitrary desire produces. (Ratzinger 1996: 230-1)

To put this point slightly differently: if one accepts, as environmentalists apparently do, that everything is interconnected and that one should adopt a holistic view of the world, then surely one cannot complain about the negative effects of human infringements of natural material laws without also commenting upon the negative effects of infringements of the moral law upon the social ecology.

Significantly, one does not have to be a Christian to arrive at this conclusion. Similar thoughts can be found in the writings of the Czech playwright and President, Václav Havel. Because of his dissident activities during the Communist era, Havel spent much time in prison. While certainly a believer in God and convinced that there is a natural law, Havel does not describe himself as a Christian. But contamination of the moral order is, according to Havel, even more insidious than environmental degradation in terms of its effects upon society. In his famous 1990 New Year's Address to the peoples of the then-Czechoslovakia, for example, Havel stated:

> We have polluted our soil, our rivers and forests, bequeathed to us by our ancestors, and we have today the most contaminated environment in Europe. Adult people in our country die earlier than in most other European countries.
>
> Allow me a little personal observation: when I flew recently to Bratislava, I found time during various discussions to look out of the plane window. I saw the industrial complex of Slovnaft chemical factory and the giant Petrzalka housing estate right behind it. The view was enough for me to understand that for decades our statesmen and political leaders did not look or did not want to look out of the windows of their airplanes. No study of statistics available to me would enable me to understand faster and better the situation into which we had gotten ourselves.

> But all this is still not the main problem. The first thing is that we live in a contaminated moral environment. (1991: 390-1)

In these words, Havel underlines his belief that humanity's primary 'ecological' problem is not to be found in the material world but *within* its moral ecology. Referring to Genesis, John Paul II makes a similar point when he states

> The dominion granted to man by the Creator is not an absolute power, nor can one speak of a freedom to 'use and misuse', or to dispose of things as one pleases. The limitation imposed from the beginning by the Creator himself and expressed symbolically by the prohibition not to 'eat of the fruit of the tree' (Gen 2: 16-17) shows clearly enough that, when it comes to the natural world, we are subject not only to biological laws, but also to moral laws, which cannot be violated with impunity. (1988: para. 34)

If, then, it is indeed true that there is a natural order that permeates our very being—and this is a powerful tradition that permeates many of the Christian churches, not to mention the writings of classical Greek philosophers such as Aristotle, Greco-Roman Stoics like Cicero, Muslim scholars such as Ibn Rushd, prominent Jewish thinkers epitomised by Moses Maimonides, and Anglo-Saxons of Edmund Burke's calibre—then one may speculate that, in this regard, Christians can legitimately direct environmentalists' attention to this dimension of the natural order.

It is also something that Christians might ask some of their co-religionists to bear in mind before the latter begin simply echoing the often profoundly materialist concerns and outlook of some environmental activists. This surely is important if Jeffrey Stout—a non-believer—is correct in suggesting that

> To gain a hearing in our culture, theology has often assumed a voice not its own and found itself merely repeating the bromides of secular intellectuals in

> transparently figurative language . . . The explanation for the eclipse of religious ethics in recent secular moral philosophy may therefore be . . . that academic theologians have increasingly given the impression of saying nothing that atheists don't already know. (1990: 110)

John Finnis agrees. If Christians, he argues, do not have anything to say, or do not want to say anything, about public issues that has not or cannot be articulated by secular humanists, then 'no-one should be surprised to find the Church ceasing to be even an interesting participant in the secular debate, and faltering in its own primary and irreplaceable purpose of leading people to salvation' (1997: 501).

Conclusion

Christianity is not a rosy abstraction. It is not a religion of escape. At the heart of Christianity lie the sinner and the humdrum mediocrity of everyday life. It commands the acceptance of the banal, the boring and the repetitive on the grounds that these too are just as much vehicles of grace as the wonders of nature, even though they are often disdained as much as the Suffering Servant in Isaiah 53. Such boring realities are often undervalued until their rhythms threaten to come to a halt. Then we see how precious, even miraculous, they are.

Thus while Christians should appreciate the natural world, they should not disdain everything in favour of nature. Christians believe that the world, as humans know it, is passing away (1 Cor 7: 31; St Irenaeus *Adversus haereses*, 5.31.1) and that the beauty of this world, however excellent, is but a foreshadowing of the next. Fortunately, more Christians are becoming conscious of the dangers posed by the flirtation of some of their co-religionists with extreme versions of environmentalism, and are willing to articulate a vision of the environment that is not only firmly grounded in orthodox Christian and Jewish thinking, but affirms that free economic activity is quite compatible with such an outlook (see Appendix).

For while environmental activism in some cases can be a worthy activity for Christians, it is no substitute for prayer, sacraments and charity. Recycling newspapers is one thing; but it can hardly be said to constitute a central feature of Christian worship. Just as Christians believe that the creation of wealth is good while the worship of wealth is simply idolatry, they should also remember that while respect for nature is good, the worship of nature is nothing less than paganism: the paganism from which many, both Christian and non-Christian, believe Judaeo-Christianity rescued much of the world.

Appendix

The Cornwall Declaration on Environmental Stewardship

The past millennium brought unprecedented improvements in human health, nutrition, and life expectancy, especially among those most blessed by political and economic liberty and advances in science and technology. At the dawn of a new millennium, the opportunity exists to build on these advances and to extend them to more of the earth's people.

At the same time, many are concerned that liberty, science, and technology are more a threat to the environment than a blessing to humanity and nature. Out of shared reverence for God and His creation and love for our neighbors, we Jews, Catholics, and Protestants, speaking for ourselves and not officially on behalf of our respective communities, joined by others of good will, and committed to justice and compassion, unite in this declaration of our common concerns, beliefs, and aspirations.

Our Concerns

Human understanding and control of natural processes empower people not only to improve the human condition but also to do great harm to each other, to the earth, and to other creatures. As concerns about the environment have grown in recent decades, the moral necessity of ecological stewardship has become increasingly clear. At the same time, however, certain misconceptions about nature and science, coupled with erroneous theological and anthropological positions, impede the advancement of a sound environmental ethic. In the midst of controversy over such matters, it is critically important to remember that while passion may energize environmental activism, it is reason—including sound theology and sound science—that must guide the decision-making process. We identify three areas of common misunderstanding:

1. Many people mistakenly view humans as principally consumers and polluters rather than producers and stewards. Consequently, they ignore our potential, as bearers of God's image, to add to the earth's abundance. The increasing realization of this potential has enabled people in societies blessed with an advanced economy not only to reduce pollution, while producing more of the goods and services responsible for the great improvements in the human condition, but also to alleviate the negative effects of much past pollution. A clean environment is a costly good; consequently, growing affluence, technological innovation, and the application of human and material capital are integral to environmental improvement. The tendency among some to oppose economic progress in the name of environmental stewardship is often sadly self-defeating.
2. Many people believe that 'nature knows best', or that the earth—untouched by human hands—is the ideal. Such romanticism leads some to deify nature or oppose human dominion over creation. Our position, informed by revelation and confirmed by reason and experience, views human stewardship that unlocks the potential in creation for all the earth's inhabitants as good. Humanity alone of all the created order is capable of developing other resources and can thus enrich creation, so it can properly be said that the human person is the most valuable resource on earth. Human life, therefore, must be cherished and allowed to flourish. The alternative—denying the possibility of beneficial human management of the earth—removes all rationale for environmental stewardship.
3. While some environmental concerns are well founded and serious, others are without foundation or greatly exaggerated. Some well-founded concerns focus on human health problems in the developing world arising from inadequate sanitation, widespread use of primitive biomass fuels like wood and dung, and primitive agricultural, industrial, and commercial practices; distorted resource consumption patterns driven by perverse economic incentives; and improper disposal of nuclear and other hazardous wastes in nations

lacking adequate regulatory and legal safeguards. Some unfounded or undue concerns include fears of destructive man-made global warming, overpopulation, and rampant species loss. The real and merely alleged problems differ in the following ways:

a. The former are proven and well understood, while the latter tend to be speculative.
b. The former are often localized, while the latter are said to be global and cataclysmic in scope.
c. The former are of concern to people in developing nations especially, while the latter are of concern mainly to environmentalists in wealthy nations.
d. The former are of high and firmly established risk to human life and health, while the latter are of very low and largely hypothetical risk.
e. Solutions proposed to the former are cost effective and maintain proven benefit, while solutions to the latter are unjustifiably costly and of dubious benefit.

Public policies to combat exaggerated risks can dangerously delay or reverse the economic development necessary to improve not only human life but also human stewardship of the environment. The poor, who are most often citizens of developing nations, are often forced to suffer longer in poverty with its attendant high rates of malnutrition, disease, and mortality; as a consequence, they are often the most injured by such misguided, though well-intended, policies.

Our Beliefs

Our common Judeo-Christian heritage teaches that the following theological and anthropological principles are the foundation of environmental stewardship:

1. God, the Creator of all things, rules over all and deserves our worship and adoration.
2. The earth, and with it all the cosmos, reveals its Creator's wisdom and is sustained and governed by His power and loving kindness.
3. Men and women were created in the image of God, given a

privileged place among creatures, and commanded to exercise stewardship over the earth. Human persons are moral agents for whom freedom is an essential condition of responsible action. Sound environmental stewardship must attend both to the demands of human well being and to a divine call for human beings to exercise caring dominion over the earth. It affirms that human well being and the integrity of creation are not only compatible but also dynamically interdependent realities.

4. God's Law—summarized in the Decalogue and the two Great Commandments (to love God and neighbor), which are written on the human heart, thus revealing His own righteous character to the human person—represents God's design for *shalom*, or peace, and is the supreme rule of all conduct, for which personal or social prejudices must not be substituted.
5. By disobeying God's Law, humankind brought on itself moral and physical corruption as well as divine condemnation in the form of a curse on the earth. Since the fall into sin people have often ignored their Creator, harmed their neighbors, and defiled the good creation.
6. God in His mercy has not abandoned sinful people or the created order but has acted throughout history to restore men and women to fellowship with Him and through their stewardship to enhance the beauty and fertility of the earth.
7. Human beings are called to be fruitful, to bring forth good things from the earth, to join with God in making provision for our temporal well being, and to enhance the beauty and fruitfulness of the rest of the earth. Our call to fruitfulness, therefore, is not contrary to but mutually complementary with our call to steward God's gifts. This call implies a serious commitment to fostering the intellectual, moral, and religious habits and practices needed for free economies and genuine care for the environment.

Our Aspirations

In light of these beliefs and concerns, we declare the following principled aspirations:

1. We aspire to a world in which human beings care wisely and humbly for all creatures, first and foremost for their fellow human beings, recognizing their proper place in the created order.
2. We aspire to a world in which objective moral principles—not personal prejudices—guide moral action.
3. We aspire to a world in which right reason (including sound theology and the careful use of scientific methods) guides the stewardship of human and ecological relationships.
4. We aspire to a world in which liberty as a condition of moral action is preferred over government-initiated management of the environment as a means to common goals.
5. We aspire to a world in which the relationships between stewardship and private property are fully appreciated, allowing people's natural incentive to care for their own property to reduce the need for collective ownership and control of resources and enterprises, and in which collective action, when deemed necessary, takes place at the most local level possible.
6. We aspire to a world in which widespread economic freedom—which is integral to private, market economies—makes sound ecological stewardship available to ever greater numbers.
7. We aspire to a world in which advancements in agriculture, industry, and commerce not only minimize pollution and transform most waste products into efficiently used resources but also improve the material conditions of life for people everywhere.

Interfaith Council for Environmental Stewardship: Serving Humanity through Faith and Reason

Post Office Box 96065
Washington, DC 20090-6065
UNITED STATES OF AMERICA
TEL: 202-628-0777
www.stewards.net

References

Allchin, A. 1978, *The World is a Wedding: Explorations in Christian Spirituality*, Darton, Longman and Todd, London.

Attfield, Robin. 1983, 'Western Traditions and Environmental Ethics', in R. Elliot and A. Gare (eds), *Environmental Philosophy*, University of Queensland Press, Brisbane: 201-30.

Augustine of Hippo, St. 1961, *De civitate Dei*, trans. R.S. Pine-Cotton, Penguin, London.

Bahro, Rudolf. 1986, *Building the Green Movement*, New Society Publishers, Philadelphia.

Bandow, Doug. 1993, *Ecology as Religion: Faith in Place of Fact*, Competitive Enterprise Institute, Washington, D.C.

Barth, Karl. 1961, *Church Dogmatics*, Bk. 3, Vol. 3, T. and T. Clark, Edinburgh.

Bauman, Zygmunt. 1992, *Intimations of Post-Modernity*, Routledge, London.

Berkowitz, Peter. 2000, 'Other People's Mothers: The Utilitarian Horrors of Peter Singer', *New Republic*, January 10: 27-37.

Block, Walter. 1999, 'Controversy: Do Market Economies Allocate Resources Optimally? A Response to Dwight Murphy', *Journal of Markets and Morality* 2 (2): 279-289.

Boff, Leonardo. 1985, *Church, Charism and Power: Liberation Theology and the Institutional Church*, SCM Press, London.

———. 1997, *Cry of the Earth, Cry of the Poor*, Orbis Press, Maryknoll.

Budiansky, Stephen. 1996, *Nature's Keepers: The New Science of Nature Management*, HarperCollins, London.

Callicott, J. Baird. 1989, *In Defense of the Land Ethic: Essays in Environmental Philosophy*, State University of New York Press, New York.

Charles S.J., Rodger. 1982, *The Social Teaching of Vatican II: Its Origins and Development. Catholic Social Ethics—A Historical and Comparative Study*, Plater Press/Ignatius Press, Oxford/ San Francisco.

Charles S.J., Rodger. 1998, *Christian Social Teaching and Witness: The Catholic Tradition from Genesis to Centesimus Annus*, vol.2, *The Modern Social Teaching—Contexts, Summaries Analysis*, Gracewing, Leominster.

Chesterton, G.K. [1923] 1987, *Saint Francis of Assisi*, Image Books, London.

Cicero 1975, *De Natura Deorum*, trans. W. Miller, Loeb Classical Library, London.

Collins, M.S.C, Paul. 1999/2000, 'Towards an Environmental Ethics', *Zadok Perspectives* 65: 9-10.

(CDF) Congregation for the Doctrine of the Faith 1986, Instruction on Christian Freedom and Liberation *Libertatis Conscientia*, St Paul Publications, Sydney.

Dailey O.S.F.S., Thomas. 1992, 'Creation and Ecology: The "Dominion" of Biblical Anthropology', *Irish Theological Quarterly* 58: 1-13.

Daniélou S.J., Jean. 1961, *Scandaleuse Vérité*, Librairie Arthème Fayard, Paris.

Denzinger, H. [1957] 1998, *Enchiridion Symbolorum, Definitionum et Declarationum de Rebus Fidei et Morum*, 26th edn., Crossroad, New York.

Dubos, Robert. 1984, *A God Within*, Charles Scribner's Sons, New York.

Duncan, Ron. 1998, 'An Optimistic View of World Food Prospects', *Agenda* 5 (1): 73-82.

Easterbrook, Gregg. 1995, *A Moment on the Earth: The Coming of Age of Environmental Optimism*, Viking, Harmondsworth.

Efron, Edith. 1984, *The Apocalyptics*, Simon and Schuster, New York.

Ehrlich, Paul. 1968, *The Population Bomb*, Ballantine Books, New York.

Ehrlich, P., and E. Wilson 1991, 'Biodiveristy Studies: Science and Policy', *Science* 253: 758-762.

Elder, David. 1996, 'The Green Utopians', in *The Churches and the Challenge of Australian Civilisation*, vol.3, *The Utopian Quest for Social Justice—Proceedings of The Galatians Group Conference*, Falcon Print, Holden Hill: 127-36.

Finnis, John. 1997, 'On the Practical Meaning of Secularism', *Notre Dame Law Review* 73 (3): 491-515.

Fox, Matthew. 1988, *The Coming of the Cosmic Christ: The Healing of Mother Earth and the Birth of a Global Renaissance*, Harper, San Francisco.

George, Robert. 1993, *Making Men Moral: Civil Liberties and Public Morality*, Clarendon Press, Oxford.

Glendon, Mary Ann. 1999, 'Rousseau and the Revolt against Reason', *First Things* 96 (9): 42-7.

Global 2000. 1980, *Report to the President*, Government Printing Office, Washington, D.C.

Gore, Al. 1993, *Earth in the Balance: Ecology and the Human Spirit*, Plume, New York.

Granberg-Michaelson, Wesley. 1992, *Redeeming the Creation—The Rio Earth Summit: Challenges for the Churches*, World Council of Churches, Geneva.

Gregg, Samuel. 1999a, *Challenging the Modern World: Karol Wojtyla/John Paul II and the Development of Catholic Social Teaching*, Lexington Books/Oxford University Press, Lanham/ Oxford.

———. 1999b *Religion and Liberty: Western Experiences, Asian Possibilities*, The Centre for Independent Studies, Sydney.

Gregg, S., and G. Preece 1999, *Christianity and Entrepreneurship: Protestant and Catholic Thoughts*, The Centre for Independent Studies, Sydney.

Griffiths, Brian. 1984, *The Creation of Wealth*, Hodder and Stoughton, London.

Grisez, Germain. 1993, *The Way of the Lord Jesus*, vol.2, *Living a Christian Life*, Franciscan Press, Quincy.

Havel, Václav. 1991, 'New Year's Address', *Open Letters: Selected Prose 1965-1990*, Faber and Faber, London: 390-96.

Hayek, Friedrich von. 1988, *The Fatal Conceit: The Errors of Socialism*, W. Bartley (ed), University of Chicago Press, Chicago.

Hodel, Doug. 1997, 'Free Markets Best Protect the Environment', *Religion and Liberty* 7 (2): 1-4.

Hore-Lacey, Ian. 1985, *Creating Common Wealth: Aspects of Public Theology in Economics*, Albatross Books, Sydney.

Irenaeus, St. 1857-1866, *Adversus haereses*, in J.P. Migne (ed), *Patrilogia Graeca*, Paris.

John XXIII [1963] 1981, Encyclical Letter *Pacem in Terris*, in C. Carlen, I.H.M. (ed), *The Papal Encyclicals*, vol. 5, McGrath Publishing, Ann Arbor.

John Paul II 1979, Apostolic Letter *Inter Sanctos*, *Acta Apostolicae Sedis* 71: 1509-10.

———. 1988, Encyclical Letter *Sollicitudo Rei Socialis*, St Paul Publications, Sydney.

———. 1991, Encyclical Letter *Centesimus Annus*, St Paul Publications, Sydney.

Johnson, Paul. 1976, *A History of Christianity*, Penguin, London.

———. 1983, *Modern Times: The World from the Twenties to the Eighties*, Harper and Row, New York.

Jones, Eric. 1991, 'The History of Nature Resource Exploitation in the Western World', *Research in Economic History*, Supplement 6: 235-52.

Leys, Simon (trans). 1997, *The Analects of Confucius*, W.W. Norton and Company, London.

Linzey, Andrew. 1976, *Animal Rights: A Christian Assessment of Man's Treatment of Animals*, SCM Press, London.

Lovelock, James. 1974, *Gaia: A New Look at the Earth*, Oxford University Press, Oxford.

Malone, C. et al. 1993, 'A Question of Fact', *Scientific American*, 269 (6): 76.

Malthus, Thomas. [1798] 1992, *An Essay on the Principle of Population as it Affects the Future Improvement of Mankind*, D. Winch (ed), Cambridge University Press, Cambridge.

Maritain, Jacques. 1970, *Three Reformers: Luther-Descartes-Rousseau*, Greenwood, Westport.

McDonagh, Sean. 1990, *The Greening of the Church*, Orbis Books, Maryknoll.

Meadows, D.H., D.L. Meadows, J. Randers, W.W. Behrens 1974, *The Limits to Growth: A Report for the Club of Rome's Project on the Predicament of Mankind*, Potomac Associates Book, New York.

Moore, Stephen and J. Simon 1999, 'The Greatest Miracle that ever was: 25 Miraculous Trends of the past 100 Years', *Policy Analysis* 364, Cato Institute, Washington, D.C.

Novak, Michael. [1982] 1991, *The Spirit of Democratic Capitalism*, Simon and Schuster, New York.

———. 1999, *On Cultivating Liberty: Reflections on Moral Ecology*, ed. B. Anderson, Rowman and Littlefield, Lanham.

Passmore, John. 1974, *Man's Responsibility for Nature*, Duckworth, London.

———. 1975, 'The Treatment of Animals', *Journal of the History of Ideas* 36: 198-218.

Paul VI [1967] 1981, Encyclical Letter *Populorum Progressio* in C. Carlen, I.H.M. (ed), *The Papal Encyclicals*, Vol. 5, McGrath Publishing, Ann Arbor.

———. 1971, Apostolic Letter *Octogesima Adveniens*, St Paul Publications, Sydney.

Pera O.P., C. ed. 1961, *S. Thomae Aquinatis De Veritate Catholicae Fidei contra Errores Infidelium (Summa contra Gentiles)*, Marietti, Turin.

Postrel, Virginia. 1990, 'The Green Road to Serfdom', *Reason*, April: 22-8.

Powell, Bernard. 1987, 'Were These America's First Ecologists?' *Journal of the West* 26: 17-25.

Primavesi, Anna. 1991, *From Apocalypse to Genesis: Ecology, Feminism and Christianity*, Fortress Press, Philadelphia.

Rad, Gerard von. 1957, *Theologie des Alten Testaments*, vol. 1, *Die Theologie der historischen Überlieferungen Israels*, Christian Kaiser Verlag, Munich.

———. 1960, *Theologie des Alten Testaments*, vol. 2, *Die Theologie der prophetischen Überlieferungen Israels*, Christian Kaiser Verlag, Munich.

Ratzinger, Joseph. 1968, *Einführung in das Christentum*, Kösel-Verlag GmbH and Co., Munich.

———. 1985, *Rapporto Sulla Fede*, Edizioni Paoline, Milan.

———. 1988, 'Consumer Materialism and Christian Hope', *Briefing* 18 (3): 43-50.

———. 1996, *Salt of the Earth: The Church at the End of the Millennium*, Ignatius Press, San Francisco.

Regan, Tom. 1990, *Earthbound: New Introductory Readings in Environmental Ethics*, Random House Press, New York.

Rousseau, Jean-Jacques. [1755] 1997, 'A Discourse on a Subject proposed by the Academy of Dijon: What is the Origin of Inequality among Men, and is it authorised by Natural Law?' in *The Social Contract and Discourses*, trans. G. Cole, Dent Books, London.

Royal, Robert. 1999, *The Virgin and the Dynamo: The Use and Abuse of Religion in Environmental Debates*, Ethics and Public Policy Center/Eerdmans, Washington, D.C./Grand Rapids.

Rydenfelt, Sven. 1984, *A Pattern for Failure*, Harcourt Brace Jovanovich, New York.

Sacks, Jonathan. 1998, *Morals and Markets*, Institute of Economic Affairs, London.

Schmitz, Kenneth. 1993, *At the Center of the Human Drama: The Philosophical Anthropology of Karol Wojtyla/Pope John Paul II*, Catholic University of America Press, Washington, D.C.

Simon, Julian. 1995, 'Population Growth Benefits the Environment', *Religion and Liberty* 5 (2): 1-3.

———.1996, *The Ultimate Resource 2*, Princeton University Press, Press.

Simon, J., and H. Kahn 1984, *The Resourceful Earth: A Response to Global 2000*, Basil Blackwell, Oxford.

Singer, Peter. 1990, *Animal Liberation*, 2nd ed., New York Review of Books, New York.

———. 1991, 'Environmental Values', in *The Environmental Challenge*, ed. I. Marsh, Longman Cheshire, Melbourne.

———. 1993, *How are We to Live? Ethics in an Age of Self-Interest*, Text Publishing, Melbourne.

Sowell, Thomas. 1983, *The Economics and Politics of Race*, William Morrow, New York.

Stein, Hannes. 1999, 'Return of the Gods', *First Things* 97: 34-8.

Stout, Jeffrey. 1990, *Ethics after Babel*, Beacon Press, Boston.

Theophilus of Antioch, St. 1857-1866, *Ad Autolycum* in J.P. Migne (ed), *Patrilogia Graeca*, Paris.

Tooley, Mark. 1999, 'Green Churches: Global Warming Among the Congregations', *Heterodoxy* 7 (5): 8.

Tuan, Yi-fu. 1968, 'Discrepancies between Environmental Attitude and Behaviour: Examples from Europe and China', *Canadian Geographer* 12: 176-191.

White, Lynn. 1967, 'The Historical Roots of our Ecological Crisis', *Science* 155 (37): 1203-1207.

Index

Religion and the Free Society:

Educating the Churches in the Principles and Processes of the Free Economy

CIS's *Religion and the Free Society* programme seeks to provide clergy, theologians, and lay church workers with a better understanding of economics, as well as an appreciation of the principles and workings of the free economy and virtuous society. In the longer term, the programme hopes to help alter the current consensus of thinking within the churches about economics, so that, at a minimum, there will be a growth in religious thinking about wealth creation and distribution. The programme seeks to achieve this end by:

• *Theological-Economic Research*. This includes building a positive corpus of religious thinking about the free economy and society, the holding of the CIS's annual Acton Lecture on Religion and Freedom, as well as occasional lectures, workshops, and seminars.

• *Educating Future Religious Leaders and Thinkers*. This occurs through the holding of Free Society conferences for those training for future ministry, theologians, church workers, etc., during which they are given an introduction to economics as well as the theological-philosophical premises of the free and virtuous society.

• *Educating Present Religious Leaders and Thinkers*. As well as organising seminars for economists, businesspersons and religious leaders, CIS produces responses to church economic statements. These responses underline any flaws in the empirical evidence or basic economic reasoning used by such documents, and, where appropriate, critique their theological and philosophical premises. Alternative arguments that provide theological support for the thinking and practices underlying the free economy and virtuous society are also expressed.

To learn more, visit our website at **www.cis.org.au**
or contact CIS on (+61) (2) 9438 4377, Fax (+61) (2) 9439 7310
If you would like to support the *Religion and the Free Society* programme,
please contact Samuel Gregg at CIS, *sgregg@cis.org.au*

Recent CIS Publications

Gambles with the Economic Constitution
The Re-Regulation of Labour in New Zealand
Wolfgang Kasper

August 2000 will see the re-regulation of New Zealand's free labour markets with the introduction of the Employment Relations Bill. Wolfgang Kasper analyses not only the Bill's likely consequences for New Zealand's economic performance—he predicts slow growth and increased unemployment—but also the risks involved in yet another change to New Zealand's ground rules of economic conduct.

[PM47] ISBN 1 8 6 4 3 2 0 4 5 0 8 (2000) 56pp

Boy Troubles
Understanding Rising Suicide, Rising Crime and Educational Failure
Jennifer Buckingham

Jennifer Buckingham throws light on the statistics of boyhood in Australia today, revealing rising crime, suicide and a decline in educational standards. She looks at the causes of these alarming changes, and offers recommendations on what might be done to deal with the consequences.

[PM46] ISBN 1 8 6 4 3 2 0 4 9 4 (2000) 102pp

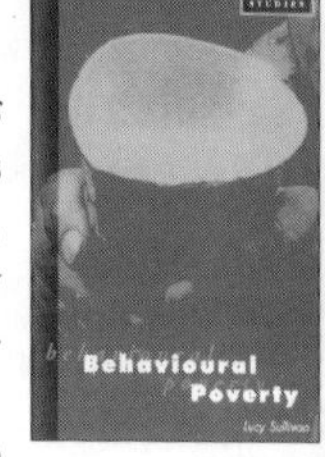

Behavioural Poverty
Lucy Sullivan

In this, essay one of *Caught in the Net: Six Essays on Welfare Systems and Family Functioning*, the author examines the welfare debate's failure to distinguish between behavioural and financial poverty. Sullivan argues against the doctrine that Utopia can be achieved by a society sharing only its money, promoting instead the rehabilitation of responsibility and competence in earning one's income.

[PM45] ISBN 1 8 6 4 3 2 0 4 7 8 (2000) 72pp

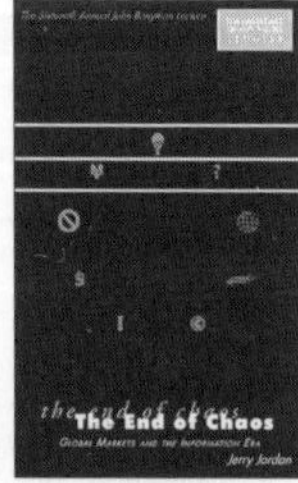

The End of Chaos: Global Markets in the Information Era
Jerry Jordan

In this, the sixteenth annual John Bonython Lecture, Jerry Jordan argues that markets will be at the heart of a new order in the 21st Century. The state needs to recognise a role in nurturing the prerequisites for the market economy, and those that fail to recognise this will find themselves discplined by global markets.

[OP72] ISBN 1 8 6 4 3 2 0 4 8 6 (1999) 20pp
